MIRACLE OF
GASEOUS STATE

MIRACLE OF GASEOUS STATE

SAEEDA BATOOL

AuthorHouse™ UK Ltd.
1663 Liberty Drive
Bloomington, IN 47403 USA
www.authorhouse.co.uk
Phone: 0800.197.4150

© 2014 Saeeda Batool. All rights reserved.

No part of this book may be reproduced, stored in
a retrieval system, or transmitted by any means
without the written permission of the author.

Published by AuthorHouse 07/08/2014

ISBN: 978-1-4969-8590-3 (sc)
ISBN: 978-1-4969-8591-0 (e)

Any people depicted in stock imagery provided by Thinkstock are models,
and such images are being used for illustrative purposes only.
Certain stock imagery © Thinkstock.

This book is printed on acid-free paper.

Because of the dynamic nature of the Internet, any web addresses or
links contained in this book may have changed since publication and
may no longer be valid. The views expressed in this work are solely those
of the author and do not necessarily reflect the views of the publisher,
and the publisher hereby disclaims any responsibility for them.

Contents

1. Inspiring Nature 1
2. Understanding Nature 3
3. Laws of Nature 6
4. Six principles of Nature 9
5. Shocking Nature 15
6. Nature and GOD 16
7. Nature as a fiction 18
8. Time 20
9. Universe 21
10. Derivation of Universe 23
11. How Is The End Of Universe? 30
12. Earth 33
13. How did human beings born 36
14. What is the reality of Death 39
15. Reality of Death of Human beings 42
16. Rays 46
17. Why is Nature Created? 47
18. What is the founder of everything? 48
19. Conclusion 49

Preface

I am Saeeda Batool ,I was born on 9th April 1995 in Quetta Pakistan .I belong to Hazara tribe ,I live in Italy and I am married. I got my education from Iqra Army Public School and Al Hadid school .I love to do something creative about science , I have wrote six articles and my first article was published on 09/4/2011 in the site of articlesbase.com and I was introduced as author. This is my first book and it means a lot to me because this is based on my own theory about nature and cosmology. There are some reasons to publish this book

To be an author ,to work and prove my theory. I have tried to prove little part of this theory with my words But I need to do a special research work on this theory this can only be done when I get a full support.

I am thankful to my Dad who teach me to walk on own my way and a special thanks to my husband who appreciated me and supported me till here. This book is dedicated firstly to Nature and secondly to my Father.

The word "miracle" means any amazing or wonderful occurrence in other hand a marvelous event manifesting a supernatural act of God. So Miracle of gaseous state means the occurrence of the first particle of gas which was a supernatural act of God. And with the occurrence of that particle of gas nature flinch with a journey, journey full of incidents, collisions, occurrence and of course surprises. Then I have listen this saying from people that "Drops gather to make a river" and here I would like to say that particles and particles gather to make bodies. Everything is amazing in nature actually nature itself is amazing, that's why everywhere we find surprises and shocks in nature. Behind every act of nature there are reasons and a complication that's why when we try to understand nature we get involve in complications. The word "life" and "Death" both have many definitions poets define them in their words, writers write them in their versions and different people have different skills about them but for me life and death is the second name of conversions. First name of them is conversions and secondly for differentiating them I am using the word bright conversion that is known as life and dark conversion which is known as death. Because conversions bring change in attitude and shape of body.

Inspiring Nature:

NATURE INSPIRES US with greenery, beauties, shocks, fictions and with its every minor move. The simplest move of nature is the small particles of gases moving everywhere in our environment. These inspirations of beauties make me love with nature and think about nature. When little particles react in between and grows in bodies, when two different powers meet and new life came in existence, when love between different relations force for great scarifies, when in a dark night stars shine, when after a dark night sun rises and a new day with new hopes begin. It's all nature which inspires us with different creations, with different means and attracts us with the lovely reflections. Inspiration makes us love with nature this is why mankind goes deeper and deeper to fulfill the dreams, the thoughts and the thirst. A philosopher fly higher and higher in the sky of knowledge to get more and more, a sea diver goes deep and deep to fulfill their thirst and a tourist goes everywhere to see the wonders of nature in all over the world. Everywhere there is a risk of life but when we fall in love with nature then there is no scar and no worries of life.

Everywhere nature inspires us with its different faces. Nature have two faces, one face of nature is visible which can be seen easily the solid nature is easy and beautiful to understand but the invisible nature is hard enough to understand like love between relations which do exists everywhere but cannot be seen, so one thing which I am interested to explain is feelings and thinking are two different things from two different part of our body. Feelings always come from heart and thinking always come from mind. This is why feelings and thinking sometimes go opposite, mostly feelings conquer thinking but sometime thinking conquers the feelings. This is just because thinking can be controlled but feelings cannot be controlled yeah if feelings and thinking is controlled that goes best.

Understanding Nature

NATURE IS TOO hard to understand yeah sometimes nature is the name of beauties but scientifically in my opinion the word nature stands for three powers in which positive and negative are opposite and must for each other and the third one is neutral, which is not must but exists in nature to control the balance between the other two charges. So this can be stated that nature circulates in a triangle of three charges.

Diagram 1
Charges

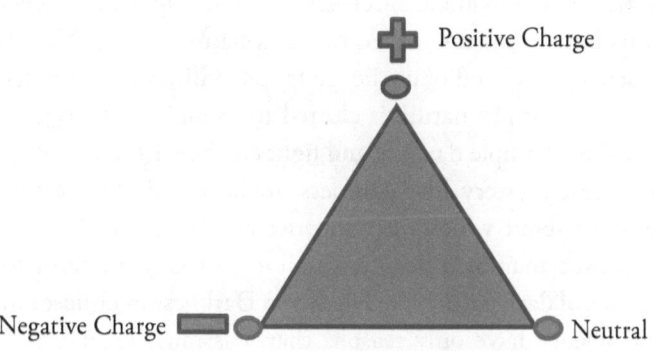

A little prove of this is that a simple atom contain three basic charges which are known as electrons, protons and neutrons. And these three charges have saved the base of nature from start now how, when and from where did these charges come from is explained further. This is a simple rule of nature that when two different and opposite powers meet each other a new life comes in creation. Everything in nature is made of two different and opposite powers either it is a small or big creation. Nature is too much spacious in different angles it gives different means for example in angle of art it gives a beautiful meaning of love in art nature means love towards everything, love towards every beauty or creation, in art the dark sides of nature is showed as a pain or scar like in seven colors of rainbow yellow is the only color which shows pain. Mankind did consider on the dark and negative nature then philosophy took birth. Then theories about nature, love, knowledge, cosmology, astronomy and every subject came forward. In which some of the theories are proved and some of them are rejected. But only a sense of art does not complete the whole nature still nature needs more explanation, practices and experiments when we did this then science took place in nature and science is the outcome of art. Nature in angle of science, scientifically nature except of its beauty has discovered the dark nature as well. Nature in science is a kind of challenge for us to discover its hidden mysteries, bright nature is cleared to us and is openly seen by us for example day, life and light etc. but the dark nature is a great mystery like darkness, night and death are that mystery about which only theories can be given but can't be proved may by a deep research or thinking we can give reasons of dark nature like Night and Darkness in philosophy and science have only reasons that it is must in nature as after several time of light or day darkness or night is must in nature. But the exact or perfect meaning of it is not proved same like this Death is one of those mystery about which only theories are putted forward is not proved. I also have a deep theory about dark nature and a little bit I have tried to prove the dark nature with my words.

These are the greatest mysteries of nature which is a kind of challenge for mankind. So scientifically nature means challenge. Well since human beings had arrive till today science have improved a lot thousands of discoveries are made but still we need more progress to discover the hidden and dark mysteries of nature we have to prove death, we have to prove soul either it exists or not. Yeah it need a deep research work but the day we discovered everything about human beings that day we will know everything about every creature of nature because nature have applied same formula everywhere in every creature.

It is hard to discover that how universe had begin? but more hardest part of discovery is that how nature has begin? For the beginning of universe we already have theories one of the theory which has been rejected was steady state theory, one of the theory which is under process to be proved is Big Bang and one of theory is mine the "Gaseous state theory" which is just a theory yet. Doesn't matter that which theory is the right one either big bang or gaseous state because soon that day will come when one of these will be proved as perfect or right one but the greatest part of discovery is to discover and prove that how nature has originate and our universe is not the first universe of nature. My theory about Derivation of universe is the first universe of nature and after that as from one human maximum 12 or 16 babies take birth same like this from that first universe uncountable universe have took birth. So to know this we should firstly understand nature then it will be easy to know how nature has begin. To understand nature let's consider on the laws and principles of nature.

Laws of Nature

NATURE IS BEING continued successfully by following some of the rules and regulations, the most important rules of nature are

1: collision of two opposite powers
2: Procedure of life in order
3: Nature Cycle

Collision of Two opposite Powers

To make a new life and continuously this process is going on from the start of Nature and will continue till the end. But what will happen if two opposite powers don't collide, if two same powers collide in nature same powers never collide a bright example is poles of magnet always north pole and south pole attracts each other north pole and north pole repels each other. Except of human beings when two same genders react in between them in these cases the output or the result is nothing. Actually all the collisions are just for proceeding further and to continue with the same process of nature and to proceed with this process we must follow the rules of nature. The three powers of nature positive, negative and neutral are everywhere in nature either it is gas, liquid or solids. In human beings also we have neutral power that is known as she males.

Procedure of life in order

Life everywhere is started with microorganisms let's take an example of start of human life. Male needs energy to produce blood to do this he eats different foods, water, fruits and milk then combining all the foods in body energy is taken to make blood. Then invisible nature works here, feelings and thinking combines to make hormones and after this procedure fixed amount of drops of blood makes sperm. Then here comes a need of opposite power. Where in female also eggs are made of energy then reaction in high temperature begins here and as a result a new life took place. Then here that new life needs a time of nine months to take a shape of body, maintain all the needs and gain a charge. After this process a new baby take birth, so this is a procedure of life in order. This procedure of life in order is not only in human beings everywhere in nature this rule is applied. Either it is universe, stars, earth, animals or human beings.

Nature Cycle

Nature everywhere is proceeding in chain or cycle we can say for example a human being is firstly in the form of gas, then it converts in liquid and after liquid it comes in a shape of body which is solid. Then after this the body grows and is alive between one century and when the body dies either it is burned, buried or whatever it converts in gaseous form. This is how nature cycle is working everywhere; everything in nature comes in gaseous form and in the end is converted in gaseous state. Everywhere I am giving the examples of human beings because it is easy to understand.

Diagram 2
Nature cycle

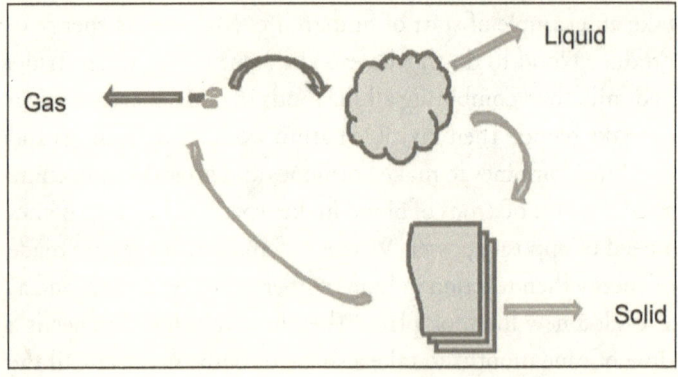

Principles of Nature: nature have some principles as everywhere in school, home, office, country or anywhere we have make systems to follow. Same like this nature involves everything universe, earth, stars etc. so this counts as a grand creation for this grand creation nature have important principles to be followed in different conditions and situations.

Six principles of Nature:

ACTUALLY IN GRAND nature there are a lot of principles in nature as nature is not something very small, nature is very much spacious so the main principles of nature which I am going to describe are the six main principles. Which normally seem totally different from each other but in invisible nature all the six principles are linked with each other and for any kind of good or bad occurrence in nature these six principles are responsible.

(1) Movement.
(2) Discipline
(3) Balance
(4) Changes
(5) Diseases
(6) Abnormalities

The very first principle of nature Movement:

The greatest principle of nature is movement. Nature is started with a move and when movement will stop in nature then its end of nature.

"The first move of a second was full of creation in which unlimited centuries were hide."

Nobody knows that when time was started and when was the first move of nature. Movement is identity of life like when an atom moves from first shell to other shells then it release or emit energy and this movement shows that the atom is alive, blood circulation of a human body is a prove of being alive, heart beats and blood moves in different parts of body, cells in a body are always moving these all are examples of movement. Same is the case with the stars; gasses blasts if the gases stop bombarding then stars will turn off and spread away in pieces. If we take a look on solids in the world so first thing which come in our mind is stone. Stones are also alive, layers in stones are always in circulation and rays are always effecting or defecting. If these movement stops in stones then it would also die. So movement is that principle of nature with which nature has took start and if movement stops everywhere then nature will end.

Discipline

Discipline also plays a great role in nature. Without discipline nature will flop, like sometimes in nature when anything gets out of discipline nature is flopped for example when discipline of circulation of gasses under world gets out of discipline earth quake occurs. Discipline is a way to follow the rules due to discipline every system is under a mannerism like our solar system. The successful system of universe is depending on discipline. Without discipline displacement will take place and destruction will cause. In the solar system discipline is must without discipline none of system can be continued successfully.

"Discipline is must for the success of any system"

Balance

Balance is also the most important principle of nature. Of course in every creation of nature amount and balance plays an important role I will like to give an example about it. May some people know this or may not earth and human body have same gasses and elements but their amounts are different. So let's come to balance, balance is must for anything. balanced diet keeps a human body perfect, balanced genetic material makes a perfect baby and so on in universe the stars and planets don't crash in between this is just because of balanced distance in between them if one of the star or planet get out of the balanced distance it will not only collide with one planet or star it can destroy the whole system. This is why balance is must in nature yeah nature sometimes flop when anything gets out of balance, so we cannot say that nature is perfect but nature can continue perfectly by following these governing principles.

These are the six most important principles of nature. In these six principles of nature five of them are must for each other movement in discipline brings a change in life and diseases are caused sometimes it disturb the life cycle but sometimes it is also good, balance govern all the four principles it keeps movement in balance, discipline and balance plays a good role in nature they both are important for each other, now changes and balance a balanced change continues the nature cycle, diseases are not friendly with balance but sometimes diseases in a balanced form are easy to control otherwise out of balance diseases are not controlled and it can broke the chain of nature. Abnormality the sixth principle of nature which is neither linked nor important for the others. Abnormality is that kind of principle in nature which is really strange to understand sometimes a question come in our mind that if there is no need of it then why it exists in nature. Now as I said before in angle of art it shows the pain this point is acceptable but only this is not enough so as I said nature is not perfect and is two sided. First of all nature is not perfect and mistakenly abnormality occurs either it is the misbalancing of sperm in ovary, out of discipline ovulation, no ovulation (female barren), a problem in sperm (male barren) etc. Nature is two sided bright and dark where nature is beautiful that is the bright side of nature but where nature comes unfavorable that is the dark side of nature. But why the dark side of nature is kept as a mystery this is just because sometimes the governing principles of nature does not work in a chain or we can say they don't link each other like I explained the chain before movement, discipline and change don't work properly and are not balanced in a mannerism in the result of which abnormality occurs, this is the dark side of nature which was kept as a mystery by nature. Prove of this is that when a baby is born abnormal by a general checkup may be in result every part of it works properly but a confusion come that basically what is the problem? Then it gets tough to find out the problem this is just because we don't know the problem from its base like at the time period of nine months when a baby is building with a body something goes wrong in that time period after taking birth and several time pass then we come to know that there is any problem with a baby then the mystery is closed and it get hard to reach till that problem.

Changes:

"Time is main the factor of changes"

Changes are that principle of nature which is linked with time and also changes do exist in nature. Time to time we come to know that how changes begin everywhere. And from the start of everything as each and everything changes physically same like that mentally or chemically changes come in appearance. Change describes time that's why in English grammar we use three words for time that are past, present and future. Now from these three words we come to know about time and we feel the time with changes. Then from the start of time till now we can see how time is changing and in every creature of nature changes work very properly. Where changes don't work properly then there abnormalities take place. Like in human when a child is grown mentally but not physically or child grows physically but not mentally they are called abnormal because changes don't works properly. So changes are also main part of nature time to time changes should occur everywhere then nature goes best.

"Nature is not changeable but its creations are changeable".

Disease

Nothing in nature is disease free. Not only human's even plants, animals and underwater world have diseases. In this way in the universe planets and stars is also not disease free. But these are other issues that diseases in them are same to these diseases may we don't know or it's not easy to understand whose diseases. For understanding these diseases I can give only one example that is easy to every one for understand it's example of earth. Like sometimes this happens that if a small part of earth is not provided water for a long time, then we provide water to that part after the long period then it will absorb the water and it may not be able to absorb water and then much difficulty would be faced due to plantation.

Abnormalities:

Nature also has defects or we can call it as a dark side of nature. Children born with abnormalities or dead many of men/women are barren. There comes a difference of gender male, female and she male. This is a part of nature. Yeah nature has defects one of the defect is abnormality sometime somewhere by mistake abnormality occurs. But maybe this is for differentiating the normal with abnormal or I think abnormality in nature is to feel the pain, pain of not having the normal thing. And abnormality is the result of imbalanced amount which occurs by mistake.

In this way, if we talk about universe. Same process of nature is going on in universe

"UNIVERSE IT SELF IS ALIVE AND EVERY MOVING PARTICLE IS ALIVE"

Stars also have abnormalities and same case is with the planets. Planets can also be barren. Some of planets may be able to produce live but some may not. Moon is a settle light which is alive but is barren…cannot produce live.

Chances of abnormalities are everywhere in nature….Even in the world we can see in humans, animals and plants….

In humans it cases in way of babies born abnormal, dead, or after a special time some of children grows physically but not mentally then they are also called abnormal…,

In the animals also these cases are include… animals also have un-creatable gender, their babies can also born dead or with abnormalities…,

In plants we can see some of trees are alive, but can't give fruits….and some of the plants are un-able for the sexual-reproduction…

Shocking Nature

NATURE SHOCKS US sometimes when something goes in inverse reaction. Rule of nature about brightness is that for brightness it is must that anything should burn like stars burn and give light, candle burn and give light, electricity burn and give light so on everything burns and give light but it shocks here when we see firefly because it does not burn and give light this is an inverse response. Nature shocks when at winter any tree gives fruit sometime, nature is shocking when after long time ovulation in some ladies occur. So this is how nature give us shock I would like to say that here nature confuse us because of its dark side and sometimes because of its imperfection. Nature shocks us with its magical creations like when we mix yogurt and cows waste in a pot and put it under sunlight for a week then after a week we will find Scorpios in that pot. Now isn't it shocking and seems like magic but actually at the other hand the reality is that chemicals in the waste of cow and yogurt reacts and forms Scorpios. Nature confuses us with its thousands of different designs that is why we think that it is difficult to reach the mysteries of nature. Because when we search on any creation it seems in a chain form like day and night is in chain process. Then should we think that life and death is also in chained form? Further I have given explanation of these questions. Let's come to this topic which is very interesting.

Nature and GOD

THIS IS A very confusing and also interesting topic firstly all over the world different people and different religions have different views about God. Some people don't believe on oneness of God but here I am not going to attack on anyone's view. I will just give my own opinion about God. Certain people who don't believe on God this is their own wish or their personal issue but the people who believe on God should first understand God. Now how to understand God is who is God? And what is God? The very first answer of the question that who is God is that God is the one who had created everything, this is simple to understand but the tough question is what is God now this is really hard to understand in some religions I have listen people saying God he, he is listening to us, he is watching us, he will be angry if we did wrong thing. This is actually wrong concept God has make the powers God is king maker, we cannot say he or she because God is not like us so we cannot give any gender to God. And we can't compare God with us this is absolutely wrong he is listening, he is watching us God does not need to watch us and God does not need to listen us. Gas, liquid and solid is God's creation it does not mean that we start comparing the creations with God that either God exist in a solid state, liquid or gas. No for example if we see a painting then it does not mean that we start thinking that the colors in a drawing are artist or brush is artist no. God is something that we can't imagine out of our thinking out of nature and out of every creation. To define God first men should define itself we cannot define our self. A human body also have two opposite powers which in a same time work oppositely I am now going to explain it in detail the body which is given to us is just temporally I cannot say that my heart is me or my brain is me or any part of my body is me. This is only my body in this body where do I really exist I don't know this I will like to give a prove about this when I am sleeping my body is laid on bed on that time when I am in my dreams and where I found myself that is me not the body laid on the bed. But I cannot find myself at the time when I am awake. As this is difficult to find our self so finding God is too much hard. Nature is God's creation which has an end but God will never end is evergreen.

Presence of God is defined as in a case we see that daily hundreds or may be thousands of people die but at the same day hundreds or may thousands of people are born, on the other hand daily a large number of deer are eaten by different carnivores animals but still population of deer doesn't finishes it remains same or get more and in sea daily large amount of small fishes are eaten by different fishes but still every generation of it do exists in sea. In these three examples I am trying to explain that in the system of life minute to minute life and creations are getting change but someone is there in the background that is controlling everything.

Nature as a fiction

Sometimes something look like a fiction but is in reality. Nature have many faces one of the invisible face of nature is emotions, every alive creature have emotions either it is water, air, sand, plants, earth, animals or human beings. Every creature has emotions water of sea when gets in relation with the full moon or weather affects it then the emotions of water rise. Weather changes and gets in relation with different things like sometimes weather gets in relation with dark black clouds and emotions of clouds and air increases and then occurs torrential rain, when air gets in relation with sand then torrential wind blows, when plants gets in relation with air, water or sand then emotions rise in them also, when earth gets in relation with sun rays then emotions also rise in earth and in result earth quake occurs. So on animals and human being are easy to understand that when their emotions rises. So this is proved and stated that when two different or opposite powers collide then emotions rises in them and new life takes birth. This is one of the most important part of nature which seems like a fiction but it actually have existence. One another face of nature is dream which have a reality but seems like a fiction, now how do dream exists one of the inner part of our body is called the main power the main power of every part of body, body is actually the hardware we can say which is temporary heart, mind or any other part of body is not the real face of us the real face of us is inside our body it is hard to define our self and I know this will seem like a fiction but actually this is reality of a human being when we sleep and see dreams, in that dreams we find our self in difficulties and we really feel that this is what that instruct us that this body is only a hardware the real software of us is invisible. Now what is that? The main power of our body is one of the master gas, which is one of the squeezed drop of all the 90 natural gases is known as the main power of any kind of alive body. Finding that gas is much hardiest part of discovery. This is why I said that universe is made of one small particle of gas having two opposite powers.

Time

TIME IS A very interesting topic about start of which many people have different theories. Time was started before the very first move of a particle; time was started when there was nothing and after a short time passed matter came in existence and again after a short time period first state of matter was created which is known as gaseous state, every creation needs a time period to maintain a body either small body or large. My opinion about time is that time is that kind of formation which has no start and will never end. Nature is created by time so nature has an end but time will never end it is like when there was nothing time was there, now there is everything among with nature time is also here and when everything will end and finish time will be there. Three words which are used for time past, present and future defines that time have no start and will never end. Time and universe is very interesting, when universe was not created what was there? There was nothing only space but time was there that is the past of universe. At that time there were other universes but I am talking about our universe.

Universe

UNIVERSE IS A kind of existence which consists of stars, planets and galaxies. Universe is being continued since Billions of years I will like to say that this continuation of universe from the start is under the rules and regulations of nature. Movement, discipline, change, diseases, abnormality and balance are working in universe to continue the nature cycle. When movement and discipline works properly then everything works in a mannered way a good example of this is our own solar system all the nine planets are rotating around the sun and in their own orbits none of planet gets out of its orbit if this does occur then the discipline will break up and displacement will take place, everything will be destroyed we think that there is only one universe but actually there are a lot of universes no one can imagine that how much universes are there in nature. There are lots of systems in universe except of our solar system. And earth is a small part of the solar system in the corner of universe. Universe is like a sky of knowledge which never ends and to define the spacious universe a human life is not enough, if a human remains alive for 10,000 000 years with a healthy life still then it is not easy to define the size and creatures of universe, only this universe which belong to us the others are so far. Because universe is much huge that we can't find the start and end of it, this doesn't mean that universe don't have start and end point it has a starting and ending point but we cannot find it. And may we end this universe and get start with new one and find different type of stars and planets may we find new type of galaxies. And life in other planets is totally opposite and different from our planet's life. This is just because nature has different faces in different places. Al l principles of nature are involve everywhere in every creature of universe, in earth and in a human body. One more interesting thing I shall share here that gases in nature are same in every creature but the thing which differs is amount of gases are different everywhere in every creature.

Diagram 3
Structure of universe

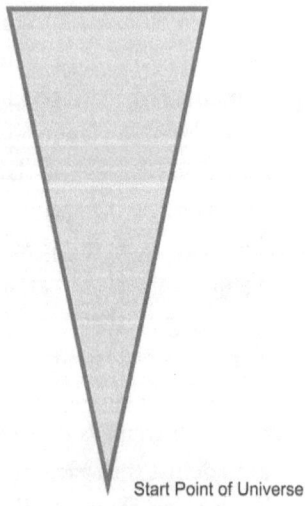

Start Point of Universe

Derivation of Universe

Now a question comes that how did the universe begin? Was there any blast? It suddenly begin or it doesn't have any start. Firstly I don't understand this that why mostly scientists think that universe begin with a blast or there Was any round shaped ball existing gases which blasted and universe started beginning. It is not necessary that grand uni has begin with blast. So for the explanation of my theory about universe are below divided in some stages.

STAGE 1:
The Gaseous state

"The Master time of occurrence in different actions"

1. Position of Master Particle (MP)
2. Master Break Down (MBD)
3. Master Maintenance of Charges (MMC)
4. Master Collision (MC)
5. Master Reactions (MR)
6. Master Birthing (MB)

In these acts the gaseous state took existence now these clauses are presenting the very first particle, gas and reactions of the origin of universe. When there was nothing let's move back up to 100 billion years ago when there was nothing everywhere was empty. Then once gently it seemed that something is there now what that was? Today's scientists have given a name of "particle" to that thing which was composed a small dotted particle at the middle point of space this is known as Position of Master particle MP1.

Now what happened next that particle of gas with no charge on it broke in two pieces it get half and both the pieces gently moved a bit far from each other till a limited distance and get in rotation this describes the Master Break Down MBD1. While this was going on the particles where getting opposite maintenance then one of the particle came up with positive charge, the other one came with negative charge on it. Charges of nature positive and negative both arise on same time and Master Maintenance of Charges was done here MMC1. After a small time both the particles started coming closer to each other. Now a time comes for Master collision MC1 of the particles with each other. They collide with each other and after the collision between them a reaction took place Master Reaction MR1. And this became a couple of gas which is named by me as "SHAZ" S1 gas means a unique gas the Father and Master gas which gave birth to all the other gases of the universe.

After less time new born particles took birth and Master Birthing MB1 was on extreme this Shaz gas gave birth to billions and billions of particles with positive, negative and neutral charges on them and every particle after birth moved in different directions as a single sign of life in those particles was movement. The particles which were moving away or roaming in empty space were known as alive and those which were not moving were known as death particles. Now as we born and after development of minimum 14 to 16 years we are able for any kind of natural reactions, same like this those new born particles also needed a time to maintain the different and opposite abilities. Now after a limited time needed reactions in the particles get started, at different time on different stages opposite particles started gathering and new type of gases took place in the grand space. May be any of that gases formed were similar to these gases which are existing now in universe. But every gas formed on that time were not these gases which are now present in universe this is because universe is being continued from long time ago and in different ages universe had changed its appearance chemically and physically. Now when new gases were taking birth which had different amount of particles and had a different Weight. This was the Gaseous state of universe.

STAGE 2:
The Liquid State

Now in stage 2 gases were reacting and movement in them came a bit fast and molecules were getting close to each other. Then there came a change in temperature also because the gaseous state was about to have a change. Soon a time came for the "Master Conversion" the gaseous state was converted in liquid state and that liquid was a strange and unique type of liquid which is also named "Shaz" S2 at that time temperature was smoothly getting high and all the gases were mixed in that liquid and the liquid needed a lot of time for the procedure of maintenance. I can confidently state this that a long time the "Shaz" liquid had kingdom in the space. In this time period liquid was having different sizes of bodies in it like molecules in water there were small round shaped bodies in that liquid. Now the main time of growth of universe was started bodies in the liquid were growing in different sizes and different colors.

STAGE 3:
The Solid State

A time comes when temperature count was getting much hot and the bodies which had ability of absorbance they started absorbing the liquid, in some bodies reactions were stopped and they died and some were not able to absorb the liquid but were alive. Because of hot temperature half of the liquid was absorb and half of that liquid remained in universe in the form of vapors and now the bodies with different amount of gases and different ability of absorbance are identified different planets and yeah bodies which did not had the ability of absorbance of liquid and had only less gases reacting constantly started burning and they are none as stars. Stars are like cells in our body. Prove of this is that the bodies which had absorb the liquid in them are planets let's take an example of our planet earth which has absorb that liquid now it have all the gases, elements and everything which is suitable for it. Similarly all the planets which have absorbed that liquid also have everything which is suitable for them in different amount. The bodies which did not have the ability to absorb the liquid and just have those gases which they are made of and are burning are stars. One more prove of this theory is that liquid was filled of that bodies which are now called planet and stars are round shaped and milky way galaxies are prove of this that liquid did exists in universe.

These were the three stages of how universe had begin.

Now this was the first universe created by nature which from start of time was expanding and this process of expanding took millions of years to reach the peak of limited place till where it was supposed to expand as I said before that identity of life in particles, atoms or any creature is movement so from each and every move of a particles or atom universe expands. And when it reached that point then the first universe of nature sprayed a mixture of gases in a small volume in the blank space. Now at one side the first universe of nature started contracting and at the other side in the space a new mixture of gases arrived and again new reactions were taking place and new procedure of life was started in same way. And then after millions of year's second universe was composed with the same formulas and same procedure but different type and color of creatures this is why if we found a universe that is totally different from our universe. And the first universe contracted and started up again with a new type of gas mixture. Now till then there were two universes but after billions of years are past till now there are thousands of universes and thousands more universes have to come up. I don't know that in which number do our universe comes.

A time came when planets started producing lives and gave birth to the different types of creatures. But some of planets did react for the procedure of life cycle but in the result they were barren cannot give life because of amount and balance of gases less, more or may the amount of needed gases were not complete. One of the example is here our natural settle light moon is also barren, some which can give life but still are unable to proceed the kind of life they had given, some gives life of same type. Like in human beings some of us are barren, some cannot give a complete live, or some are born dead and some give same type of life (only male or female). These cases of abnormalities are everywhere in nature either they are human, planets or universe. And the main part of beginning of universe is that every state of matter from the start is taking round in orbits.

Diagram 4
Expanding of universe

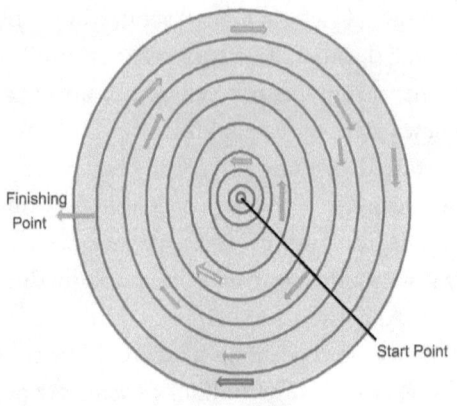

Today scientists are conforming that the universe is expanding and expanding is the growth of universe but tomorrow scientists will accept that the universe is contracting because every action have reaction this is action of universe that it is expanding and in reaction it will contract like our breath inhaling and exhaling, day and night and weather. As the time taken by our breath is smaller than the time of day and night and time of action and reaction of day and night is smaller than time of weather. This is why universe expanding take much longer time of centuries to complete the action going and one thing more till this action of expanding is going on there will an increase in the amount of gases till the universe will not reach its limited place and when it will reach the exact place then a little amount of mixture of gases will be sprayed in blank space, where there is nothing. And after that universe will start contracting and reaction will start on then the amount of gases will start decreasing and when universe will fully contract on the point from where it took start then the life of universe depends on temperature if temperature will be full high then gases may vanish off but if the temperature will be not much hot then may new reactions take place and universe will start on with new type of life. Then life will change in new shapes. Like when we exhale then lots of gases are taken out of our body and when we inhale new and fresh gases are taken into our body.

Diagram 5
Contraction of universe

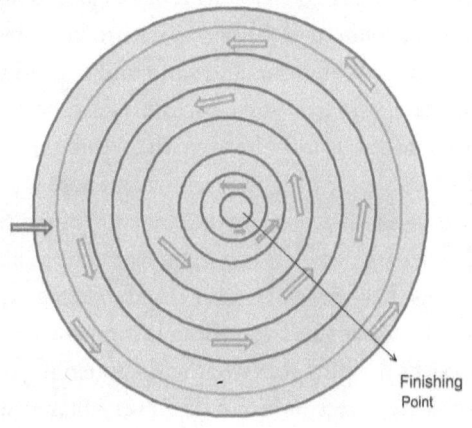

Finishing Point

HOW IS THE END OF UNIVERSE?

Everything which has a start it is must to be end. So the universe which has taken a start it has to be end up one day. Now how will it end? Everything in nature has a base, earth have three element water, fire and sand and it is said that our base is from sand and it is proved while a dead body has been buried for six or seven months after that time the dead body is dissolved completely in sand. So human's base is sand when human life is end it dissolves in sand. So every creature of earth will dissolve in earth when it end, now when the conception of universe comes that how will universe end so let's go to the base of universe which was gas. Everything is composed of gas states of matter so let's imagine that what will happen if gas flesh out from everywhere. Lives on earth and every planet will finish and force of gravity will finish from everywhere and everything will be ruin earth, planets, star and galaxies. So this is what will happen for the end of universe, the main power which has made universe, because of which universe is continuing and because of which universe will end is gas. Solid, liquid and plasma states of matter are made of gas. But one thing more I shell explain that earth is a part of solar system and solar system is a part of universe earth and other planets will not end with universe because life of universe is more than life of planets. So if any planet finish its life it will break in pieces and spread away in different parts of universe and this does not effect on any other bodies of universe because each and every body in universe have a limited distance in them which keeps a balance between stars and planets.

Now one thing differs the meaning of end of universe, as before I said that when universe starts with life then it start expanding and will expand till a limited place. When it reaches the peak point it spray a mixture of gas in empty space and will start contracting. Now when universe expands the amount of gas in it gets increasing and when it reach the point where it stops expanding the amount of gas gets full and this is why a small quantity of gas is dropped down or sprayed in space. After this when universe starts contracting then gases in it gets decreasing. At the time of contracting universe start getting small and this may not take a lot of time. Then may suddenly each body in universe start dying, there will be change in temperature and every solid body will start melting again liquid state of matter will take place and after that liquid will convert in gaseous state and all the gases will dissolve in one single gas particle now this depends on defensive power the one with great power will dissolve all the other gases in it. If that gas starting reactions in it then there will be a chance of life again a new universe will take birth this will not be end of universe this is just end of the lives in universe then same process of beginning of universe will again start with new type of gases and lives. End of universe is actually that time when the single particle of gas will finish.

This is sure that one day universe have to be end, universe will end and nothing will left there and this will only happen if the small atom or particles of gases started dying then every minor and huge creations will start ending up. And this process of ending may take a little time and may there cause an inverse reaction temperature will be high every solid creature will start melting in a liquid form and after all the solids will convert in a liquid form then from liquid state it will convert in gaseous state in a form of vapors and then all the gases will start dying everything will be finished again matter may compose. Now this composition of matter may finish of or it may compose new gas and again this procedure of life take a new start.

Diagram 6
Universes in space

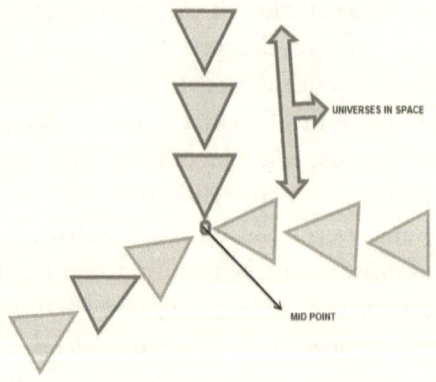

This was a brief theory about start and end of universe.

Earth

From a deep thinking it is stated that human is descendant of earth. Earth is combination of two opposite powers. Nature applies same principles on every creatures of it, like male and female reacts together to produce new life, like day and night reacts together for continue life, etc. Now this is how earth is producing life with reference to gap in timings. Earth do react with sunlight to produce life, rays in between earth and sun forms energy in earth and several types of lives take birth.

Now when the process of beginning of universe had been done, universe has begin and conditionally is ready to lead a life with all of its bodies, every bodies of it are alive and may some of them are miscarriage, dead or abnormal. So let's think about life of earth, earth was one of those planets which were born save and healthy. After earth was born it took a period of time to maintain its body level in this time period all the gases, liquid (water) and solids were completed. Now firstly water the specific liquid of earth was already in earth before its birth at the time when earth was absorbing that liquid of universe from that time water was made in earth. Gases and liquid were already in earth the term completed means that they were reacting perfectly and were on the right position, ok now about solids. One of the solid materials is sand which was made at that time when earth was under process but at that time sand was not dry it was wet and May sticky then they get dry because of change in temperature. Earth itself was hot because of lavas circulating inside it but time to time earth also prepared cold. Another thing stones and mountains are surely made of lava when under world gases were reacting in between them and sometimes it goes in imbalance way when gases crush in between them in the result of which earth quack occurred and lavas came out from earth in high pressure and were cold down because of cold temperature and are now known as mountains. And one more thing lavas under world are always revolving like our blood circulation. After these procedures first creation of earth came in creation now what was that first creature of earth? A strange question that what really was the very first creature of earth so let's come to the point that when gases of sand and water reacted between them then plants took birth and this reaction was going on since the start till now, everywhere on earth plants took birth in this way and plants were spread everywhere it is not important that the first plant on earth was similar to these plants which are now on earth surely their shapes were different and one more thing that their properties of energy were also different just because of them the second type of creature took birth and that second type of earths creature were insects which were born from plants in different ways like some were born underground in the roots of plants, some were born on the plant like on the trees and some were born from the fungus when a fruit or vegetables were expire then they cause a fungus in this case some insects were born. Insects are that type of creature which is born if somewhere has fungus and if somewhere is sand and water. Now the third important creature of earth is animal, how an animal did took birth on earth. As this is scientifically proved that before the arrival of human beings on earth there was a kingdom of dinosaurs and before dinosaurs there were any other creation and before them any other I don't know how many creations where there before dinosaurs so let's talk about the third type of creature on earth which were animal and animals were outcome of insects now I will like to explain this process with this saying "every incident gives birth to a new type of life" from the day earth starts its journey of life incidents and incidents were occurring one prove of this is that water and sand are two opposite or I shell say different elements of earth now one of the incident is this that when they reacted in between them plants took birth and when plants took birth then again an incident opposite properties of plants reacted with sun light and sand then insects took birth and when opposite insects reacted with each other then small sizes of animals came in creation so on when these animals and insects reacted for the reproduction in between them and opposite powers and properties mixed with each other then different countless types of animals and other creations where increasing day by day and so on underwater world was also going on with this process and earths scenery was also getting different day by day. Now a day came when a creature took birth in large size and that were known as dinosaurs, dinosaurs were spread in all over the world by having the reaction with opposite couples in countless times. Now what happened for the shutdown of dinosaurs? There came a climatic change temperature of earth begin very cold all over the world and in the result of which dinosaurs started dying.

How did human beings born

DIFFERENT PEOPLE HAVE different theories about how did mankind arrived on earth. Beside all theories I will give a brief explanation on my theory about the arrival of mankind on earth. As before it is stated that human beings are not the first creation of earth so the process of life and death was going on before the arrival of mankind. Before us here in the earth was a kingdom of dinosaurs and how they died can surly be answered that they died because of climatic crises. And after their shut down did not shut down for a short period imagine maximum of one year there was nothing alive on earth except of the gases. This was a kind of rest for earth because all the gases were changing in new shapes we can say that every quality of each and every gas changed. And after the rest time when earth again maintained the ability of producing life then again different reactions took place. And incidents happen when sunlight rays give energy to earth and gases started reactions then, water and sand chemicals reacted in the result of which plants took birth, different plants reacted and population of plants increased. Then secondly small insects took birth in different places like in the roots, on the plant and inside the fruits. So on reactions were going on quickly and incidents on incidents were happening. In different and opposite sexual reproductions were taking place, where the amount and energy of gases were high there the creature were powerful but where the amount and energy level were low there the creature were with less power or either small in size. But where the amount of gases were high and energy were low then the creature were large in size and low of energy, somewhere amount of gas were less but energy was high in that case the animal was small in size but had a lot of energy. In this way the population of earth increased time to time.

And a time came for a grand incident which gave birth to the "Homo Species" now an incident happened here were two dissimilar animals opposite or different in every quality rate of energy and amount of gases were totally different and opposite reacted sexually and what happened that female animals are familiar with producing numerous eggs at a time. When they reacted sexually then at a time numerous eggs where fertilized with different genetic material and chromosomes. Now as the animals where not of same generation for example monkey and chimpanzee did this. In the result of which a new type of creature took birth that was Homo habilis. Now at a time numerous homos took birth may two, three, four or more than this took birth. As these babies were child of that animal so she the female one did the custody of them. She feed them her milk and so on time to time the Homo habilis were maintaining their bodies and get older. The females one and the males one were maintaining their different and opposite abilities and qualities. Now a after 14, 16 or may 17 years when both the genders were ready for the reproduction process they did reacted sexually and these reactions proceed further and further several times and again new babies took birth. So on these processes were going on with different kind of changes like change in appearance and mind capacity because of the difference in kind of energy and gases. Then came the second generation of human that is known as Homo erectus. With change in its every quality and proceeding further brings that generation which belongs to us Homo sapiens.

This was a short history of spacious life the Homo species.

What is the reality of Death

WITH DIFFERENT THEORIES different people says that when human is died then after seven re-cycling life in different faces man meets God, some says that when we are died then we are made star, some says that we are taken to another planet and some says that after our death we will remain in other world till the judgment day.

"Death is nothing except of changing shape of life in Gaseous form."

We are always confused about death of course death is the dark nature about which only theories are put forward. But is not proved because to prove anything we need to do a practical work on it and we need to experience it, death is that reality which cannot be experienced. So this is why I have a kind of theory which doesn't need experience. Death is changing shape of life in the form of gaseous state. Let's understand the reasons of death. Different creatures in nature have different life time ability first of all the small particles of gases when they are born and till virgin have different qualities in them but when they get in reaction with other particles the qualities in them get change then again with some more reactions they get more change in them this is how life change in new shapes in gaseous state. Now let's come to universe as universe from the beginning is expanding and this expanding goes on in it till a limited place and sprays a mixture of gases in blank space there in a blank space a new type of life of universe is started and on the other hand the death process of the first universe gets start after that it will start contracting and when this contraction reaches the exact point then all the gases is dissolved in one, two or may be three types of gases then this is death of one universe but creation of new universe with different and new type of gases. This is how the life and death cycle of grand universes continues. Same like this stars and planets also have death cycles. Now in nature different bodies have different amount of gases, where the amount of gas is less there the cycles are small but where the amount of gases are more there the cycles are long. As small particles or atoms we can say have fewer amounts of gases they have very small cycles. I mean that they change their shape of life in seconds. Then comes stars now stars also have lots of gases in them but here in them differs the movement of gases if the movements of gases in them are normal they are perfect but the ones without normal rate of movement dies earlier. Because if movement in one star is higher it losses all its energy quickly that is why it turns off and spreads away in pieces in universe, the one with low movement gases can't recover with time that is why it turns off and spreads away in pieces. Then what happens two things may happen firstly may any body near to any piece pull that piece of star towards it and with any kind of reaction it dissolves the piece of star. Or the second option is that either the pieces of star remain in universe for any other reaction. Now the death cycle of planets, planets remain alive till the gases are reacting in them, movement in them are normal from movement I mean to say the circulation of mixture of gases inside the planets like lava inside the earth. Now this can easily be understood that every ability of earth is continuing on the bases of gas for example the gravitational force, the magnetic force and rotating force. If gases flesh of inside and outside the earth every force will finish, everything will destroy and earth itself will break down. So in planets the life time of gases and energy differs when a planet dies it also break in pieces and spread away in different parts of universe.

Reality of Death of Human beings

Diagram 7
Reality of Death

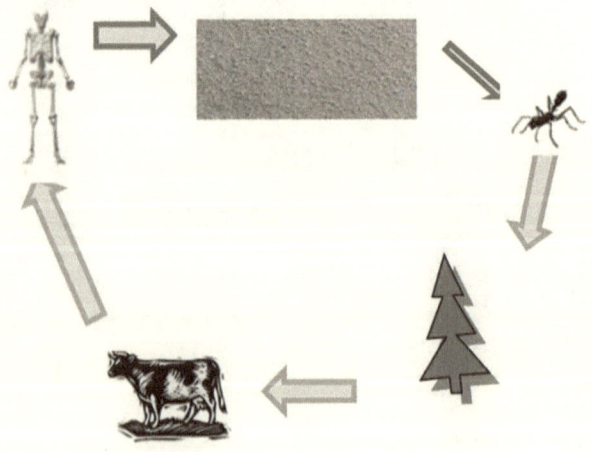

Death in life on earth means conversions the exchange in form and type of life. This is going on from the start of life on earth till now, because every creature of earth has a limited life time different in times. And this depends on several things but the most important one is energy. Energy plays a great role in every kind of life, like when a human being is child has less energy that is why diseases affect the kids easily. After it energy level is higher in adults that why diseases cannot attack adults more easily and the older one also have less energy. But energy is must continue in same speed we can say. Because if the energy level gets down then every part of body will slow down and then there comes chances of death. Death has three main reasons.

1 Accidental death
2 Diseases Attack
3 Defensive Powers

In these three cases always death occurs now what happen after death with the dyed body which are buried Sand and the dyed body effects each other and reactions take place in the result of which again energy is produced and is divided in small organism that are insects then that insects start eating the dyed body and gains more energy now energy of a dyed body is converted in that insects. After this as small species have small amount of energy so they also die and gases are changed here so when the bodies of insects are mixed with moisture then energy is converted in the form of plants. And now if the plant is eaten by any animal then energy goes in animal and after it if that animal is eaten by human then energy goes in human and after several stages again when human reacts sexually then energy comes again in the form of human. So death means conversions of energy in different attitudes.

If a human is dyed in accident any kind of accident which also includes bomb blast body pieces are spread in pieces then some of the energy move in air in different directions. And energy moves in any kind of body to be fertilized in the form of breath. And gets fertilized there and again comes in the form of body.

If a dyed body is burned then energy converts in the form of charcoal sand and when that charcoal sand is mixed with water then energy change its form and is manufactured in water for any kind of other creature.

Now this energy which I am talking about is easy to understand this is that energy which always remains in our body and which keeps us alive. And is in a high amount in our body in daily life we may waste a lot of our energy but it is maintained again by eating or drinking. As the energy wasted by us also remains here in nature and by lots of conversions is bring back to us like a good example is daily we may burn lots of papers and that burned pieces of papers are thrown in any place that pieces may mix with moisture sand or water and take existence in the form of tree. With the wood of that tree may again papers get manufactured and supply till us. As at the starting I said that life and death is the second name of conversions.

This is how life and death cycles work daily and converts in seconds

End of Human Life

Without any reason nothing is done, same is the case here we the creature who call their self-human beings are the most superior organism of earth, but actually we are still not as superior as we think. In this modern age where use of technology is common we cannot compete nature we are not struggling hard for humanism in this age of technologies our fellow being are facing a lot of problems even we can't create an environment of brother hood. Doesn't matter if diseases are naturally with us but in a short lifetime we can spend a good life. Because for us it is Horner that nature has gifted us the life of human being hence in nature every moving particle is alive, but nature has given us the authority to spend a good life. And make use of its every thing.

And this is persuaded that human life which was started from end of the other life and this has to be ended one day, may there come any crises, may there would be a natural disaster or man made things defect in a bad manner and human life come in a crises for example 'Nuclear power (atom bomb)' it is extra energy, there is no advantage of it in use of human it only have disadvantages. Because human is that supremacy which can compromise every problems like climatic crises, change in nature etc. some misbalancing, defects and erroneous can cause a big problem for every creature of earth at the same time. And every creature of earth will come in crises nothing will left plants, animals and human will vanish off from the world. This end of human will give a birth to a new creature of earth but if the end of human beings will naturally occur then there will be not much loss, if man made things defect in a bad manner and any kind of disaster did occur by a mankind then we are surly gone but after us earth will face a lot of problem. If gases stopped reactions then earth will remain barren till long time, till the gases recover or if the gases failed to recover then earth will be fully barren.

But it will be good if gases recover again and start again with reactions, incidents and life cycles. Then every creature will be in new shape. Again earth will give birth to the other creations. And this is the principle of nature. Not only earth is following this principle of nature nine other planets are also there under the system of nature. Which are also following this principle of nature.

Rays

RAYS ARE BRAND new technologies discovered by us the human beings, but still it need more discovery till now settle light system and some more systems like mobile signals, remotes etc. works with rays but a new revolution will come in world when we will discover the technology of rays hundred present. Then everything will work on rays. Rays are the important source of nature; energy to every living creature of earth is supplied by sun to earth through rays. Rays are the reason of the every creature which is born on earth. When sun rays fall on earth and reacts with every part of earth then daily thousands of creatures are born and thousands of creatures die. Because rays are always effecting on earth and its every creature. Sometimes I had listen this that the solar eclipse and lunar eclipse do effect on a pregnant women that are the rays effecting on a pregnant women, rays of solar eclipse badly harm eye side. Human body always emits rays and the other rays are always effecting on a human body. In this 20th century the rays which are discovered are the quarter part of rays the other rays technology have to be discovered still we need more progress. We need more progress to proceed further and develop more to reach the peak of technology.

Why is Nature Created?

WHY IS NATURE created is a tough question which is not easy to understand. Philosophically it can be answered that nature is created for joy. But scientifically this cannot be proved because only joy is not the complete reason of creation of nature. What did actually happen in the result of which the first particle of nature came in existence? Keeping the dark sides of nature under view I can say that nature has begin randomly. Because if the reason of beginning of nature is just for joy then why abnormalities, diseases, pain and dark faces do exists in nature. Nature is not only the name of beauties it also have draw backs. But the sense of nature has been created randomly does not defines the first particle because this cannot be proved that first particle of space has took birth by its own. Beside every theories the first particle of nature was been created prove of this is that when first particle came in existence then it was a little particle and it needed a lot of requirements and conditions it was passed from lot of processes after which it was divided in two parts. And the two master gases where given the required conditions, keeping the principles and laws of nature under view I am stating that Nature has a creator. It has not begin itself.

What is the founder of everything?

THIS IS AN easy question that what is the founder of everything all over the nature. Nature means universe, stars, planets, earth, human beings and every creature. So founder of everything is nature and nature is created by gas. Gas is the founder of everything, but what is the founder of gas? Founder of gas is mystery of nature. Either the creator has putted a dot of gas or anything else did happen in the result of which the shaz gas was born and continued further.

Conclusion

Here comes the binding up of the book Miracle of gaseous state as I have explained God and its every supernatural act. From the base nature took place with only one supernatural and amazing occurrence when a partial gas particle was formed. And then it passed from different conversions and after a lot of conversions a universe was formed and this was the first universe which was formed in three different stages. When the whole universe was formed then also life was materializing everywhere and then planets and stars the curved bodies took place in different attitude. Then comes our own planet earth which was luckily alive planet then after certain development again luckily produced life. After different survivals of life in earth a time came when the origin of Homo creature took birth by a couple of different animals. Then this creature started a journey of generations and till now we are the superior and amazing creature of earth because we have a large volume of mind capacity. Reality of death actually is the conversion of energy from different stages to different forms. Then in the transformation of energy rays play an important role, this transformation of energy is in chained form for example:

Plant ➔ Animal ➔ Human ➔ Sand ➔ Insects ➔ Plants ➔ Other forms.

This is how life circulates in the form of energy, in this chain firstly plant is eaten by animal then animal is eaten by human when human is dyed then energy is converted in sand then insects are born and after this energy of all the insects are converted in plant then in end other forms of energy means that may that plant is eaten by human or may by animal and again the journey of life is started through different types of conversions. Rays are the amazing source of energy in nature from different means, everywhere we see the effects and defects of rays clearly that how rays are effecting our daily life. This is not necessary that nature in its every act goes perfect but where nature moves imperfect there it turns in new lives because nature is spacious. In past years I always think that if in daily life we lose 50% of energy then where from the other energy comes from like if we burn trees then after some time again new trees are born. When I considered on nature then I understood that if daily we lose energy then we also gain energy from minor to huge bodies.

www.ingramcontent.com/pod-product-compliance
Lightning Source LLC
Chambersburg PA
CBHW021037180526
45163CB00005B/2165